Earthquakes

Illustrations: Janet Moneymaker
Design/Editing: Marjie Bassler

Earthquakes
ISBN 978-1-950415-36-6

Published by Gravitas Publications Inc.
Imprint: Real Science-4-Kids
www.gravitaspublications.com
www.realscience4kids.com

Have you ever been in an **earthquake?**

If so, you may have felt the ground shake.

I think I do **not** want to be in an earthquake!

Or you might even have seen buildings rock or fall down.

But what is an earthquake?

I wonder.

? ?

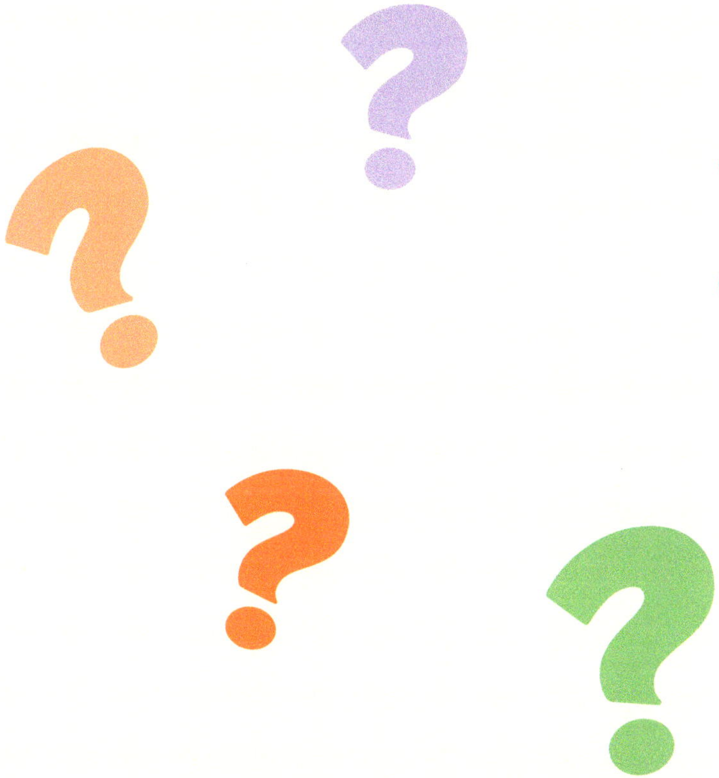

Recall that Earth is made of **layers.**

Earth is made of different layers.

The outer layer is the **crust.**

The middle layer is the **mantle.**

The innermost layer is the **core.**

The **crust** is solid.

The **mantle** is solid on top and soft like peanut butter below.

The **core** is soft metal on the outside and solid metal in the middle.

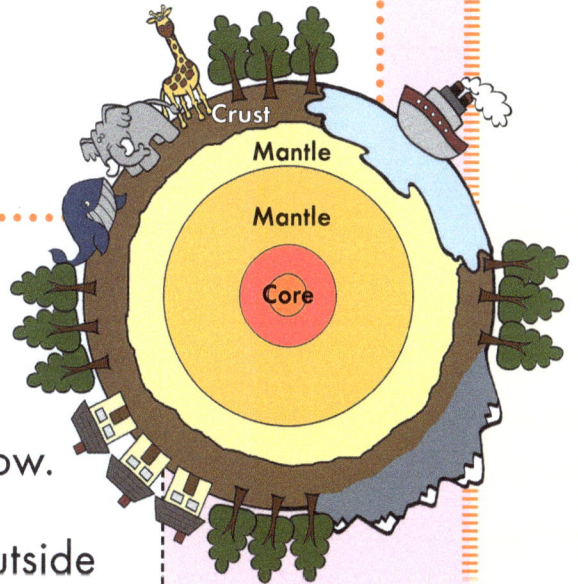

Also recall that Earth's crust and solid upper mantle are broken up into huge pieces called **plates.** The plates fit together like a puzzle.

The place where the edges of two plates meet is called a **fault.**

Surface of Earth

Plate

Plate

Plate

Plate

Crust and
upper mantle
(solid)

Lower mantle
(soft magma)

Magma
moves

Edges of plates

Earthquakes can occur where the edges of Earth's plates meet.

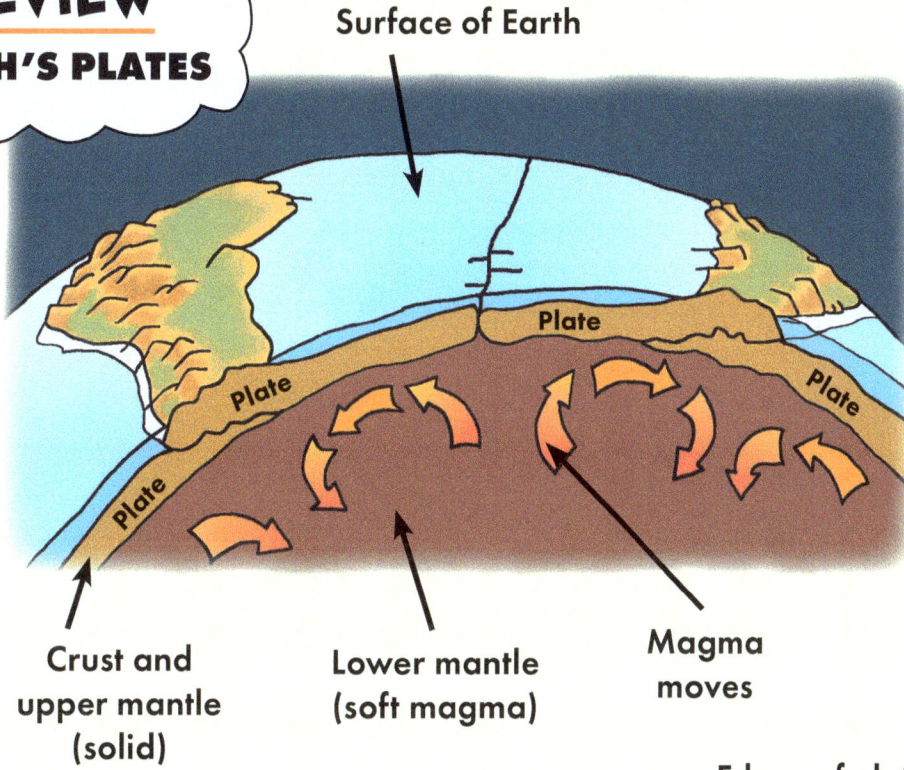

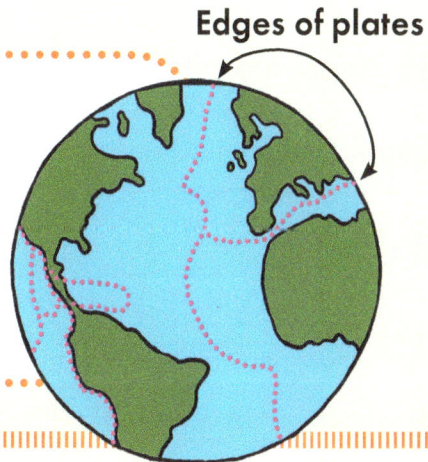

The soft materials in the lower mantle move. As they move, they carry the solid plates of the upper mantle along with them.

> >

Edges of plates meet (faults)

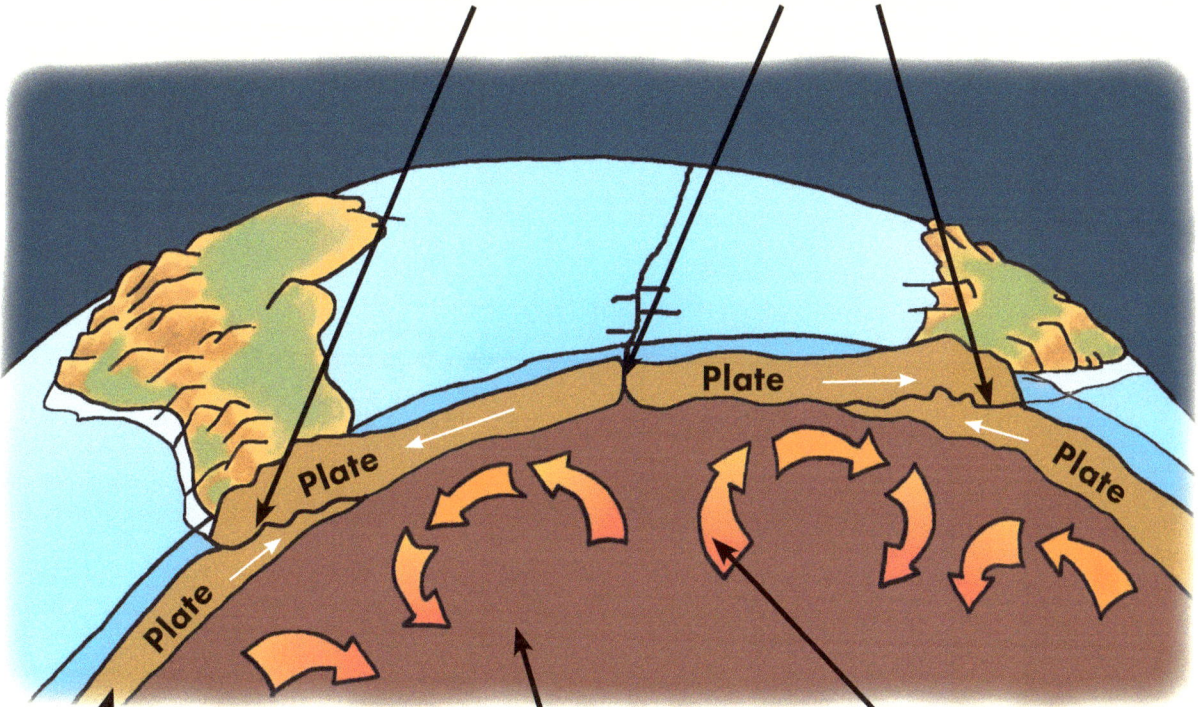

Plate

Plate

Plate

Plate

Crust and
upper mantle
(solid)

Lower mantle
(soft magma)

Magma moves,
carrying the
plates with it

The plates move, but the rough, rocky edges of the two plates stick together at the fault.

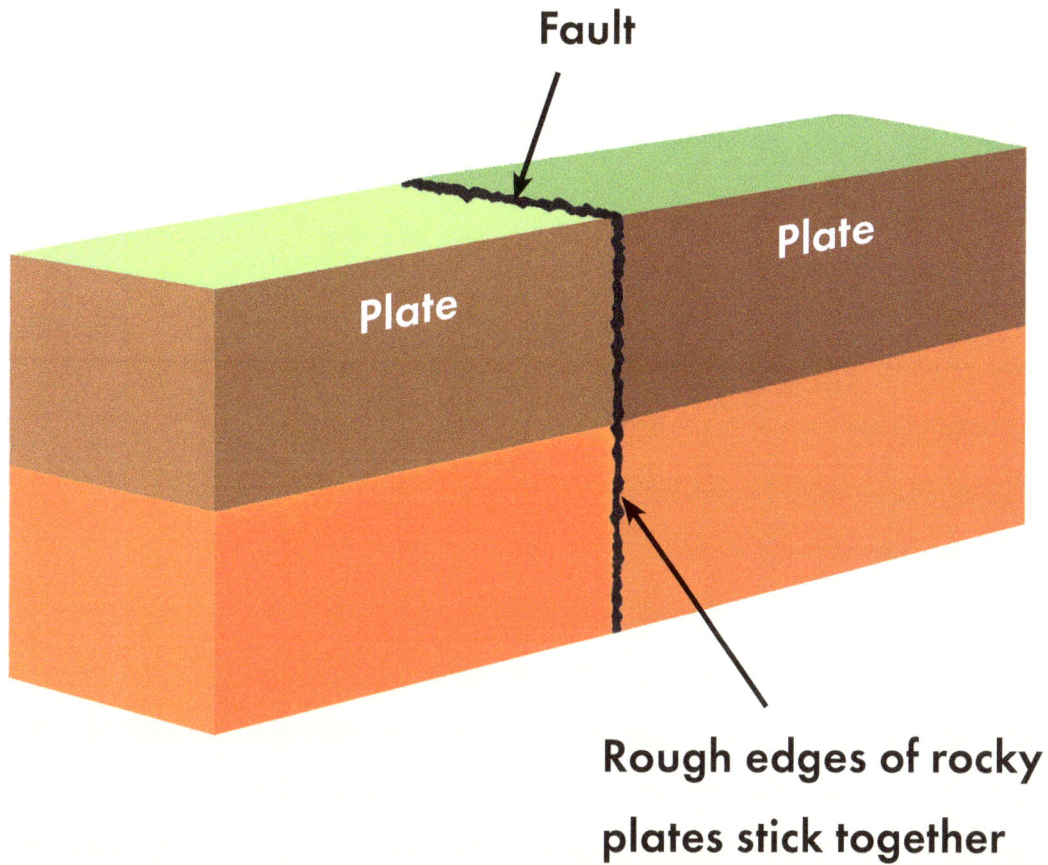

Illustration adapted from domdomegg, CC BY SA 4.0

Finally, the edges of the plates can no longer hold on to each other. They suddenly let go. This sudden movement of the ground causes an earthquake.

EARTHQUAKE!!!

EARTHQUAKE

The edges of the rocky layers suddenly let go

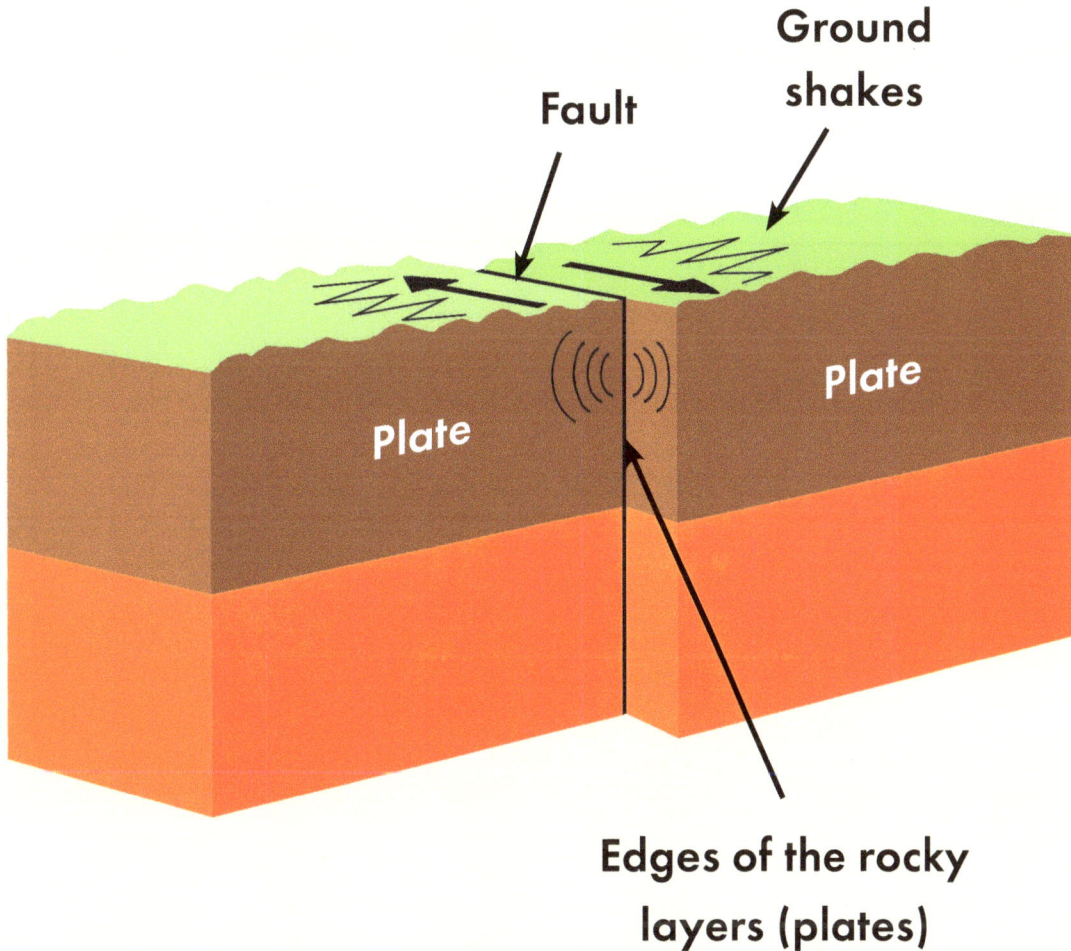

Fault

Ground shakes

Plate

Plate

Edges of the rocky layers (plates)

Earthquakes create **seismic waves**.
Seismic waves are vibrations that transfer
energy through the Earth from the source of
the earthquake.

We can feel big earthquakes far away from
their source because we feel the vibration of
the seismic waves.

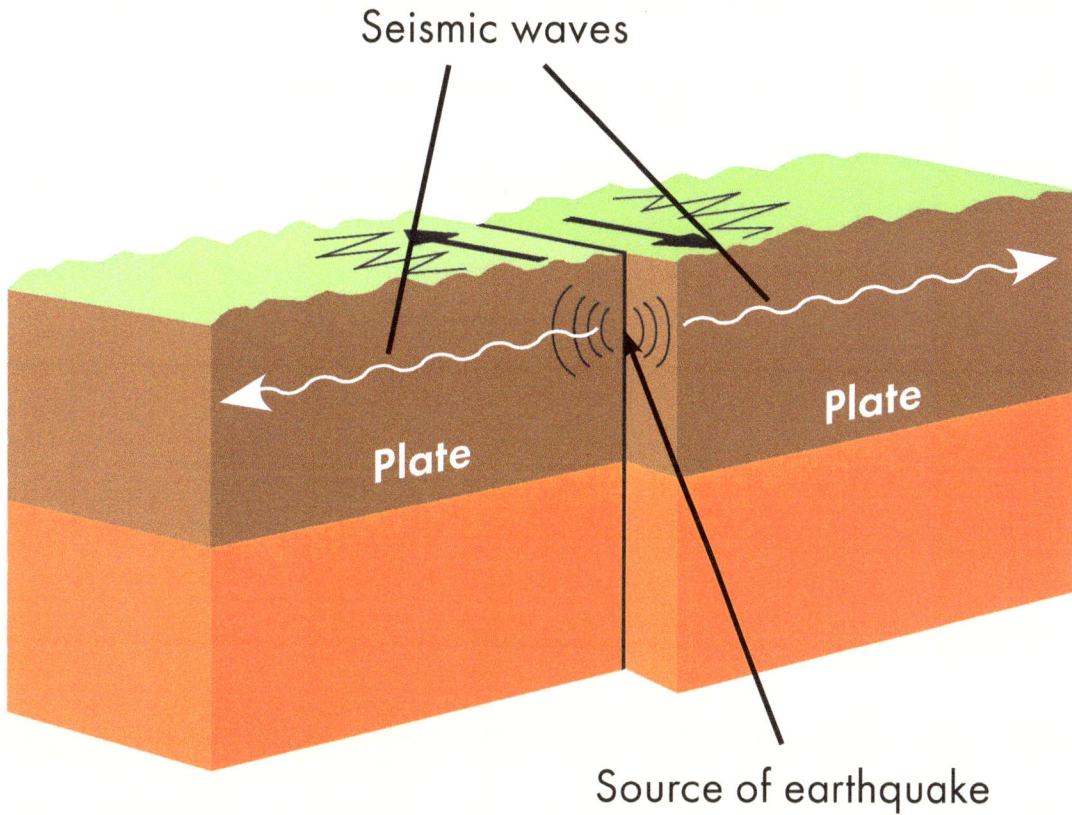

Seismic waves

Plate

Plate

Source of earthquake

A big earthquake can cause a lot of damage. Many other earthquakes are so small that people do not feel them.

By studying earthquakes, geologists can learn a lot about how Earth works and what the inside of Earth is like.

Learning about earthquakes is fun!

How to say science words

core (KAWR)

crust (KRUHST)

earthquake (ERTH-kwayk)

fault (FAWLT)

geologist (jee-AH-luh-jist)

layer (LAY-uhr)

magma (MAG-muh)

mantle (MAN-tuhl)

plate (PLAYT)

science (SIY-ens)

seismic (SIYZ-mik)

wave (WAYV)

What questions do you have about EARTHQUAKES?

Learn More Real Science!

**Complete science curricula
from Real Science-4-Kids**

Focus On Series

Unit study for elementary and middle school levels

Chemistry
Biology
Physics
Geology
Astronomy

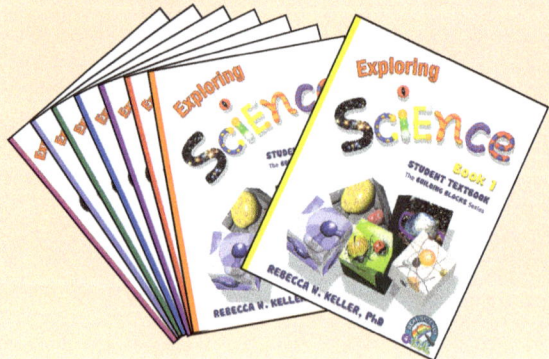

Exploring Science Series

Graded series for levels K–8. Each book contains 4 chapters of:

Chemistry
Biology
Physics
Geology
Astronomy

www.ingramcontent.com/pod-product-compliance
Lightning Source LLC
Chambersburg PA
CBHW040149200326

41520CB00028B/7546